Pre-Kindergarten Jumbo Workbook:
Little Math Specialists

SPEEDY
PUBLISHING

Speedy Publishing LLC
40 E. Main St. #1156
Newark, DE 19711
www.speedypublishing.com

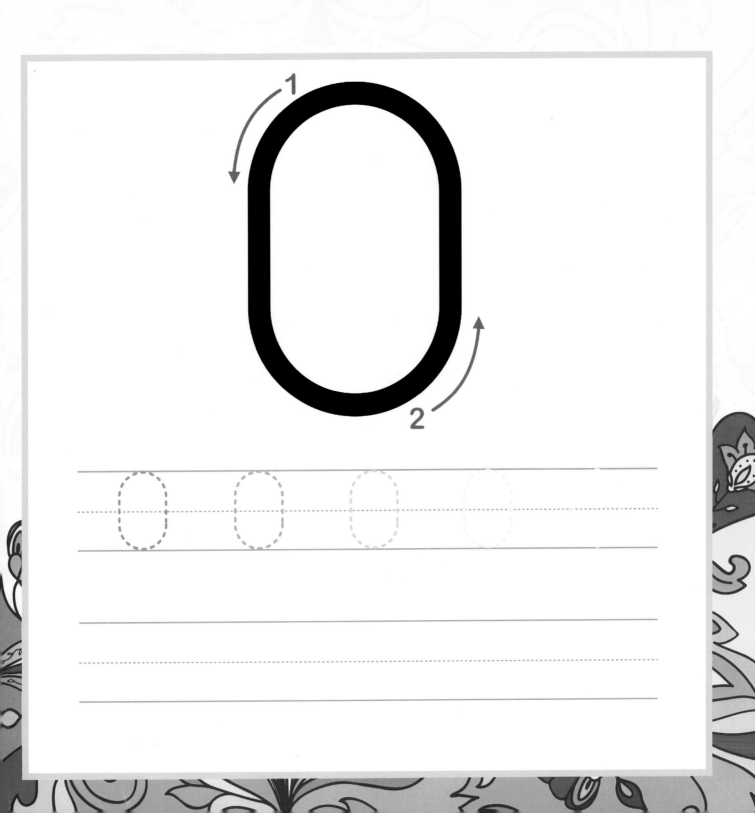

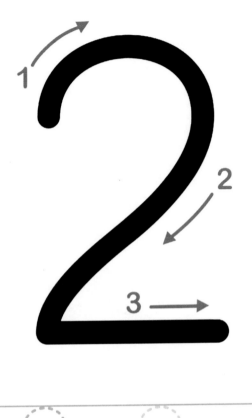

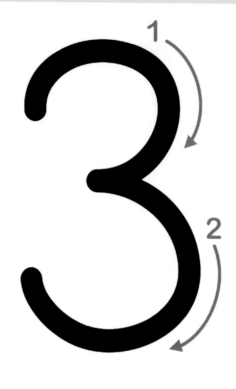

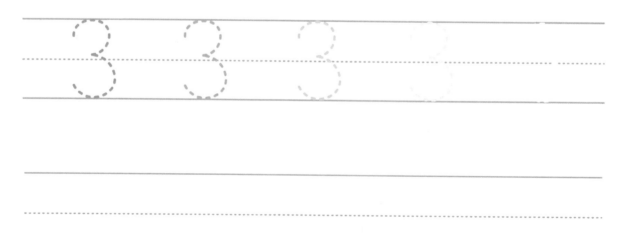

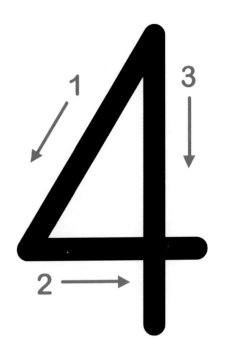

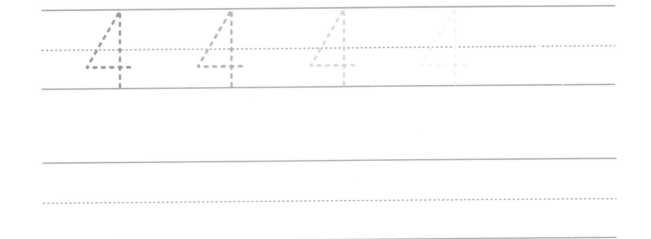

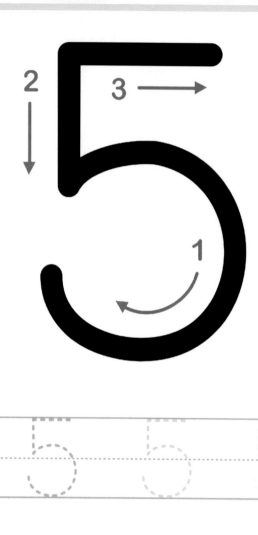

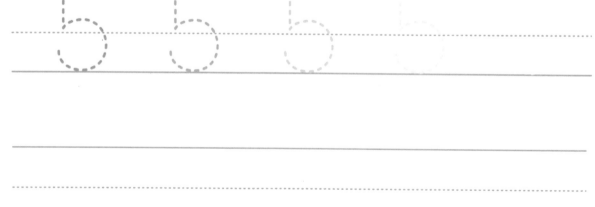

Let's Count!

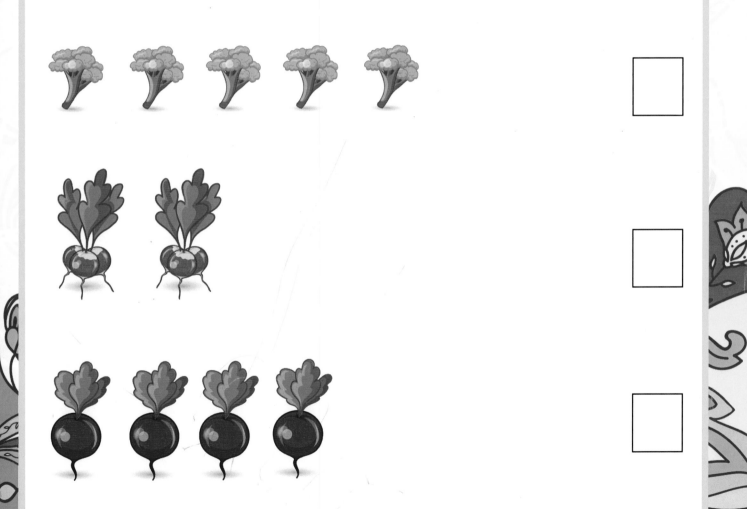

Complete the Numbers!

___ 1 2

___ 4 5

___ 7 8 ___

HOW MANY SHEEP DO YOU SEE?

?

HOW MANY BUNNIES DO YOU SEE?

HOW MANY

ELEPHANTS

DO YOU SEE?

?

HOW MANY

ROBOTS

DO YOU SEE?

HOW MANY SANTAS DO YOU SEE?

HOW MANY

BIRDS

DO YOU SEE?

HOW
MANY
DOGS
DO YOU SEE?

?

HOW MANY
PIGS
DO YOU SEE?

HOW MANY

FISH
DO YOU SEE?

HOW MANY COWS DO YOU SEE?

HOW MANY ALIENS DO YOU SEE?

?

HOW MANY

BEARS

DO YOU SEE?

Connect each set.

 • 4

 • 5

 • 3

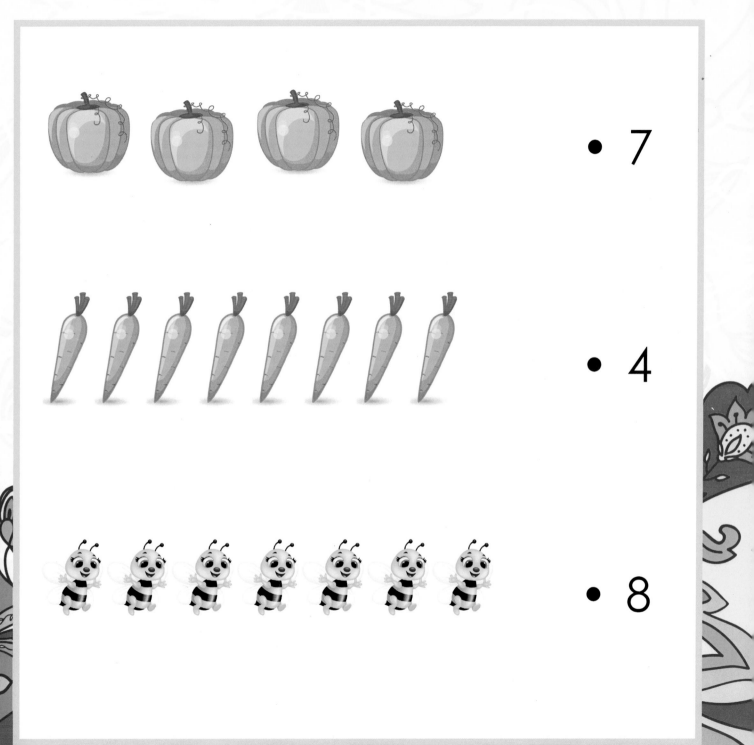

• 7

• 4

• 8

 4

• 5

• 8

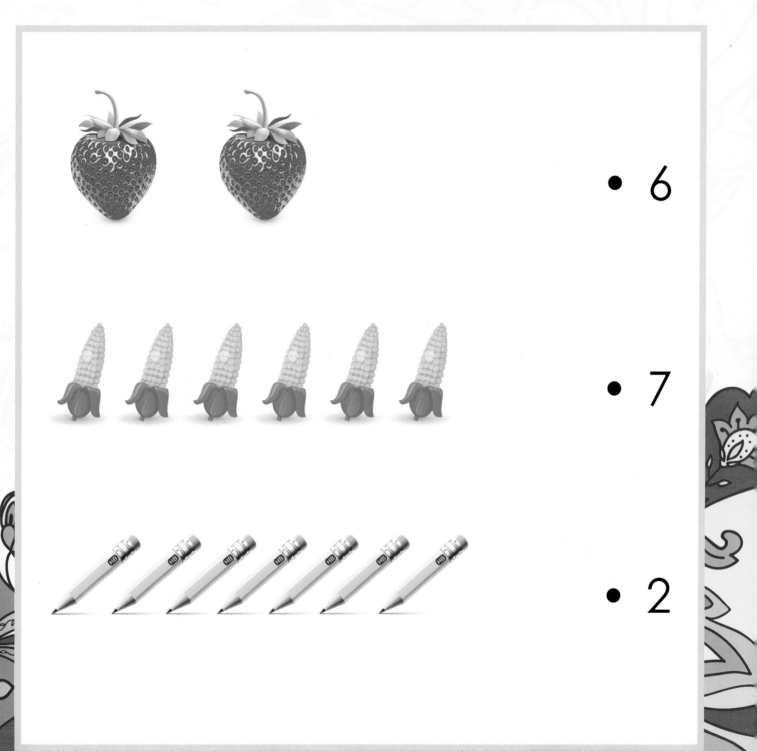